See for Yourself

Rain

See for Yourself

Rain

Kay Davies and Wendy Oldfield
Photographs by Robert Pickett

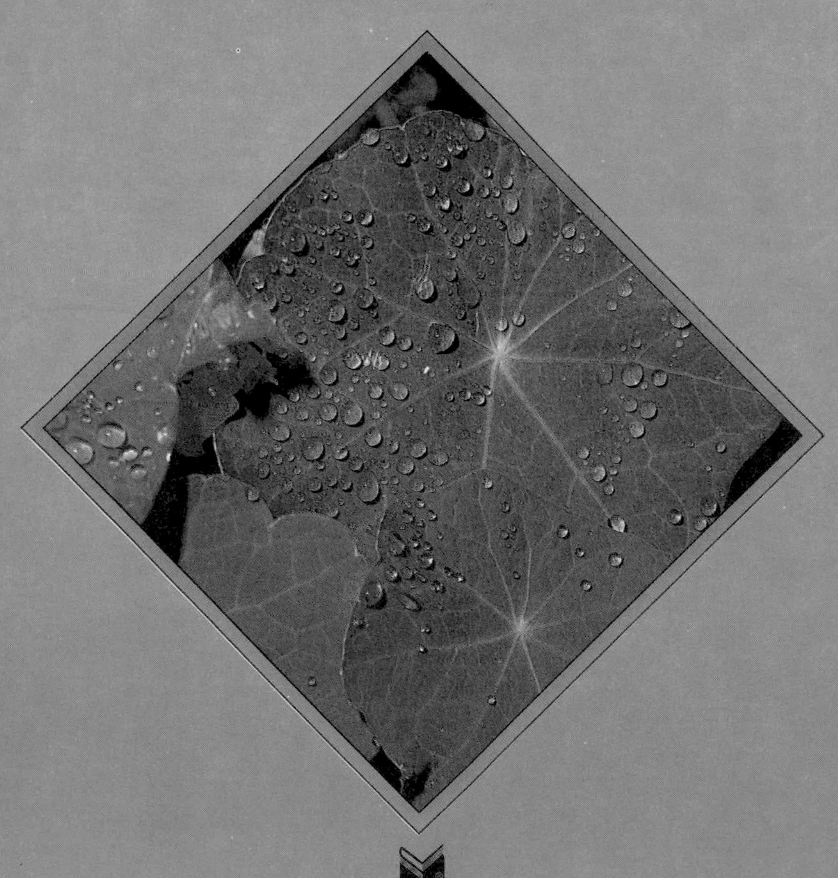

RAINTREE
STECK-VAUGHN
PUBLISHERS
The Steck-Vaughn Company

Austin, Texas

Published by Raintree Steck-Vaughn Publishers, an imprint of Steck-Vaughn Company

Editor: Kathy DeVico
Project Manager: Julie Klaus
Electronic Production: Scott Melcer

Library of Congress Cataloging-in-Publication Data

Davies, Kay.
Rain / Kay Davies and Wendy Oldfield; photographs by Robert Pickett.
p. cm. — (See for yourself)
Includes index.
ISBN 0-8172-4043-8
1. Rain and rainfall—Juvenile literature. 2. Water—Juvenile literature. 3. Water—Experiments—Juvenile literature. [1. Rain and rainfall—Experiments. 2. Water—Experiments. 3. Experiments.] I. Oldfield, Wendy. II. Pickett, Robert, ill. III. Title. IV. Series.
QC924.7.D38 1996
551.57'7—dc20 95-6005
CIP
AC

Printed and bound in the United States
1 2 3 4 5 6 7 8 9 0 99 98 97 96 95

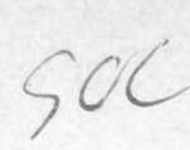

Contents

Everything Gets Wet!

Do you like going out in the rain?
Look at the big picture. Can you see the rain making puddles on the sidewalk and running into the drain?

If you are out in the rain, the water falling on your hair and skin can make you feel cold and wet. Your clothes and shoes may soak up water, too. What can you wear to keep yourself dry in rainy weather?

Do you have any of these things?

Do you know what they are made of?

How Birds Stay Dry

Do you know how birds stay dry in the rain? The bird in the big picture is called an avocet. It is covering its feathers with oil from its body. Oil makes birds' feathers waterproof. Drops of water cannot stick to their oily feathers.

Human skin is waterproof, too. Your skin makes oil to keep water out.

Try dropping water from a medicine dropper onto your bare arm.

What happens to the drops?

Round Raindrops

Look at the picture of raindrops on a leaf. Most of them are round. The surface of water acts like a skin. It pulls each raindrop into a smooth, round shape.

Try watching water drip from a faucet. At first, the drip is shaped like a pear. But once it has fallen from the faucet, it pulls itself into a ball again.

The insect in this picture is called a back swimmer.

The back swimmer hangs upside down from the surface of the water. This is because the surface of water is like a skin.

Water Vapor

Rain makes puddles. Have you ever noticed how puddles shrink and disappear on a sunny day?

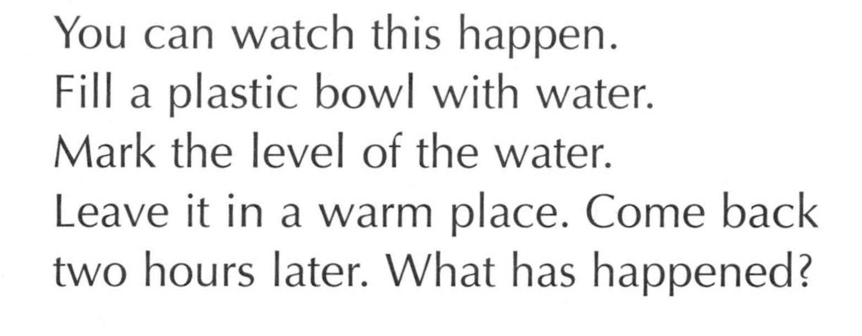

You can watch this happen. Fill a plastic bowl with water. Mark the level of the water. Leave it in a warm place. Come back two hours later. What has happened?

When water gets warm, it turns into a gas, called water vapor. Water vapor rises into the air. You can't see it. But you know when water has become vapor, because the water disappears.

This marsh in the big picture was once full of puddles. But now all the water has turned into water vapor.

Clouds Make Rain

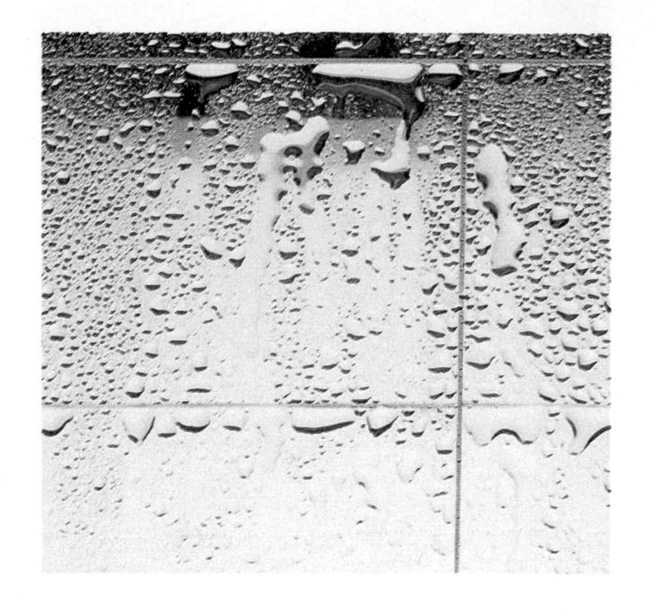

Water vapor mixes with the air. When it cools down, it turns back into drops of water. Have you ever seen this happen in a room inside your house?

(Try the bathroom!)

When water vapor in the air outside turns to water, it can form clouds.

Clouds are made up of very tiny droplets of water that float in the air. Sometimes the tiny droplets may join together. They become bigger, heavier drops. These fall to the ground as drops of rain.

Look at the big picture. Can you see the rain falling from the clouds?

From Solid to Liquid

Have you ever noticed what happens if you put sugar in your drink? The sugar disappears. It mixes with the water.

Try putting some sugar cubes into a cup of warm water. Watch them melt away.

Look at the big picture of a cave. Water has turned the rocks into liquid. Strange shapes have been left behind. This is because the water has dripped from the rocks and then turned into vapor.

This happens very, very slowly. It takes thousands of years!

Rain Fills Rivers

When rain falls, some of it soaks into the ground. Some of it stays on the surface. Rainwater collects on high ground and helps to make rivers. Can you see that the river in the picture is running downhill?

You can find out what shape a river makes as it flows downhill. Fill a plastic tray with damp sand. Tilt one end. Gently pour water from a watering can onto the top end of the tray.

Why do you think your river spreads out as it reaches the bottom end of the tray?

Rivers Work Machinery

Fast-flowing water has a lot of power.
Look at this picture of a waterwheel.
You can see that the flowing water strikes the blades. This pushes the wheel around.

The turning wheel drives the machinery inside the mill. Inside this mill, the machinery grinds wheat into flour.

You can see how this works. Pour water from a pitcher onto a plastic waterwheel. Can you make the wheel turn quickly? Or more slowly?

Can you make a toy that uses water power?

Plants Need Rain

How many different kinds of flowers can you see in the big picture? All plants need water. They can only grow if they are watered.

Plants grow roots. Their roots grow downward. They suck up water from the soil. The water goes up their stems and into their leaves.

You can watch this happen. Color some water with food coloring or ink. Put a carnation in the colored water. How long does it take for the water to reach its petals?

Animals Need Rain

The zebras in this picture are drinking from a lake. The lake is full of rainwater. All animals need water. Without water, they would die.

Do you know how many glasses of water you drink in a day? You could count them. Fruit juice is mostly made up of water. You can count this, too.

Can you think of any other drinks that have water in them? Look at the picture below.

On the Lookout for Rain

The cloud in the big picture is a special shape. This kind of cloud will almost always bring heavy rain.

Some things will show you when it is going to rain. Stand a pine cone in a window. It will open up when the weather is dry, like the one in this picture. It will close up again just before rain.

Find some seaweed, and hang it outside. You will see that it gets damp if rain is on the way.

Try watching the sky. What other kinds of clouds usually mean that it is going to rain?

More Things to Do

1. Look through a drop of water.
Use a medicine dropper to place a drop of water onto a piece of plastic wrap. Place it over a square piece of paper. Look through the drop. What do you notice? Put your water drop over some printed words, too. What happens?

2. Sayings about rain.
What do you think these sayings mean?

It's raining cats and dogs!
When it rains, it pours.
To save something for a rainy day.

3. Make a rainbow.
Put a glass full of water in a sunny window. Place a sheet of paper on the floor below the window. Move the glass around until you see a rainbow on the paper. Can you name all the different colors of a rainbow?

Index

This index will help you find some
of the important words in this book.

Notes for Parents and Teachers

As you share this book with children, these notes will help you explain the scientific concepts behind the different activities, and suggest other activities you might like to try with them.

Pages 6–7
Boots are made from rubber, which keeps water out entirely. Today's waterproof clothing is made from fabric that has holes in it that are too small to let water in, but large enough to let water vapor (sweat) out. Umbrellas are made of finely woven fabric, which lets air and light through, but not water droplets.

Pages 8–9
The sebaceous glands in the hair follicles produce an oily substance, called sebum. This lubricates the skin and makes it waterproof. When water is poured onto the skin, it will roll off in drops. Children should be encouraged to notice the texture of their skin when they stay in the bath for a long time. It will get dry and wrinkly, because the sebum oil has been washed off.

Pages 10–11
Water comes out of the faucet as a smooth round stream or as pear-shaped drops, because surface tension pulls it inward. Surface tension enables the back swimmer to hang upside down from the water's surface.

Pages 12–15
Evaporation and condensation are the two key processes in the water cycle. Water evaporates from oceans and rivers, rises into the air as water vapor, condenses into liquid water, and falls as rain. Rain soaks into the soil, waters plants, and fills rivers. Eventually, it returns to the ocean to complete the cycle and start again.

Note: It takes approximately one million cloud droplets to form one drop of rain.

Pages 16–17
Rainwater dissolves limestone more easily than any other kind of rock. Rainwater is a very weak solution of carbonic acid, which can dissolve rocks made of calcium or magnesium carbonate. In limestone areas, rivers hollow out potholes and underground caves. Stalactites hang from the ceiling of these caves, and stalagmites grow from the floors. These long spears of rock are formed as water drips and evaporates, leaving a little of the dissolved limestone behind.

Pages 26–27
Children can watch for the different kinds of cloud formations: cirrus clouds, which are thin and wispy; cumulus clouds, which are puffy, like cotton balls; and status clouds, which form in layers. White cumulus clouds often mean showers, and a heavy buildup of the cumulus cloud type indicates bad weather is on its way.

The cloud in the picture is a cumulonimbus cloud. It is a storm cloud, which could bring thunder and lightning, hail, and/or rain.